MicNic **Editions**

Text and illustrations : Mickaël Nicotera

Nature Logbook

Mickaël Nicotera

This logbook belongs to

....

*Stick here a bird feather
or a dry leaf*

Trees

You can easily identify a tree simply by the kinds of leaves it produces.

The margin of a leaf is an important and characteristic feature. It can be sinuous, toothed, rounded, lobulated, ...

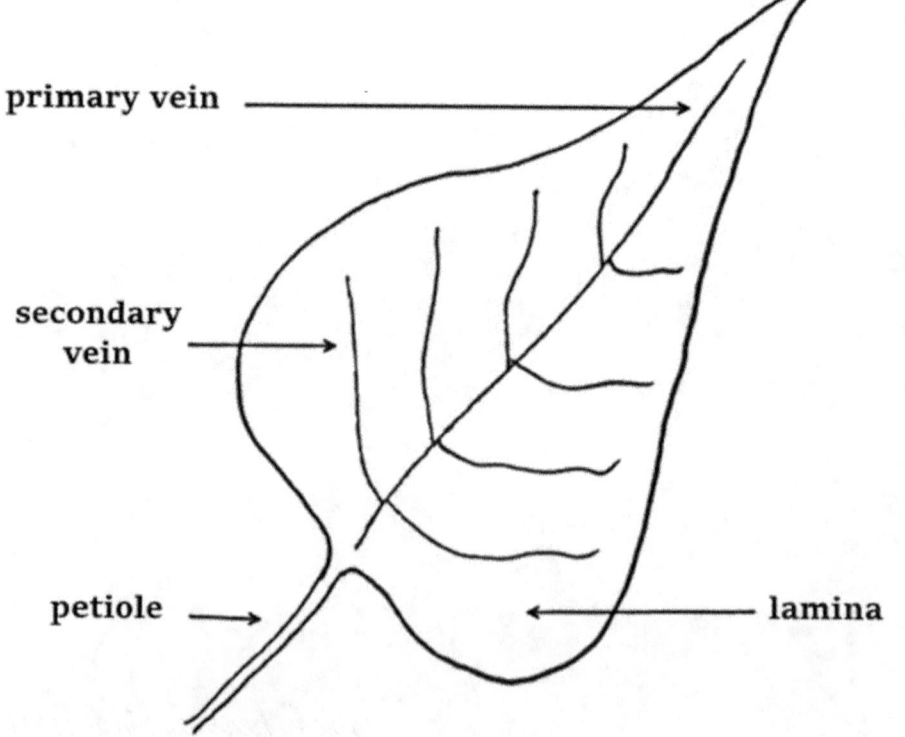

Common trees

Oak	Beech	Hornbeam	Birch
Alder	Hazel	Elm	Mountain nash
Sycamore	Hedge maple	Maple	Chestnut
Horse chestnut	White poplar	Black poplar	Willow
Apple tree	Pine	Cedar	Spruce

Tree observation sheet

Probable name :

Date :

Leaf diagram :

- ☐ acuminate
- ☐ auriculate
- ☐ serrate
- ☐ cordate
- ☐ crenate
- ☐ deltoid
- ☐ palmate

- ☐ rounded
- ☐ falcate
- ☐ sinuate
- ☐ tumcate
- ☐ pinnate
- ☐ lobed

Color(s) :

© All rights reserved – Nature logbook by Mickaël Nicotera

Tree observation sheet

Probable name :

Date :

Leaf diagram :

- ☐ acuminate
- ☐ auriculate
- ☐ serrate
- ☐ cordate
- ☐ crenate
- ☐ deltoid
- ☐ palmate

- ☐ rounded
- ☐ falcate
- ☐ sinuate
- ☐ turncate
- ☐ pinnate
- ☐ lobed

Color(s) :

© All rights reserved – Nature logbook by Mickaël Nicotera

Tree observation sheet

Probable name :

Date :

Leaf diagram :

- ☐ acuminate
- ☐ auriculate
- ☐ serrate
- ☐ cordate
- ☐ crenate
- ☐ deltoid
- ☐ palmate

- ☐ rounded
- ☐ falcate
- ☐ sinuate
- ☐ turncate
- ☐ pinnate
- ☐ lobed

Color(s) :

© All rights reserved – Nature logbook by Mickaël Nicotera

Tree observation sheet

Probable name :

Date :

Leaf diagram :

- ☐ acuminate
- ☐ auriculate
- ☐ serrate
- ☐ cordate
- ☐ crenate
- ☐ deltoid
- ☐ palmate

- ☐ rounded
- ☐ falcate
- ☐ sinuate
- ☐ tumcate
- ☐ pinnate
- ☐ lobed

Color(s) :

© All rights reserved – Nature logbook by Mickaël Nicotera

Tree observation sheet

Probable name :

Date :

Leaf diagram :

- ☐ acuminate
- ☐ auriculate
- ☐ serrate
- ☐ cordate
- ☐ crenate
- ☐ deltoid
- ☐ palmate

- ☐ rounded
- ☐ falcate
- ☐ sinuate
- ☐ turncate
- ☐ pinnate
- ☐ lobed

Color(s) :

© All rights reserved – Nature logbook by Mickaël Nicotera

Birds

Birds can be identified with their colored plumage and the shape of their beaks. Males are often much more colorful birds.

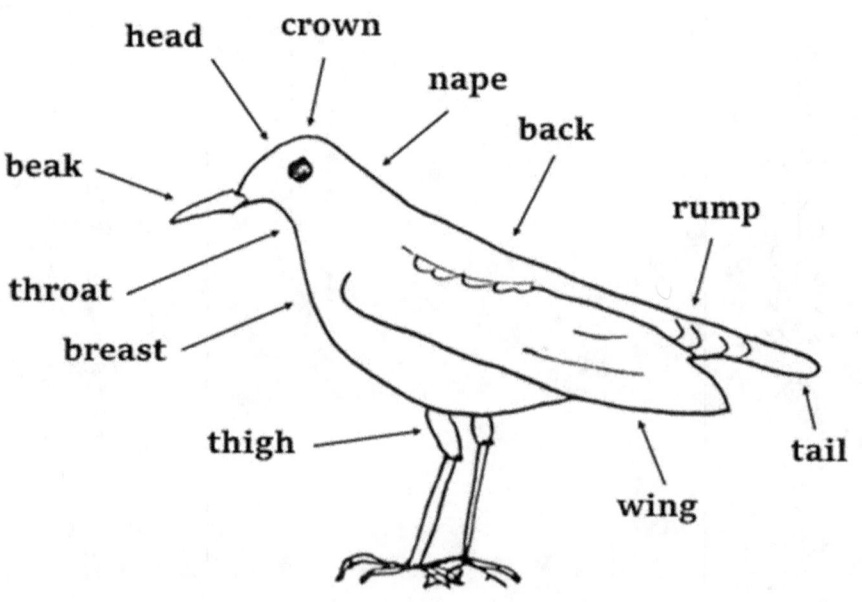

Bird species

Insect catching	Grain eating	Scything	Generalist
Surface skimming	Probing	Nectar feeding	Carnivore (Rapace)
Coniferous-seed eating	Pursuit fishing	Dip netting	Fruit eating
	Aerial fishing	Filter feeding	

Bird observation sheet

Beak :

☐ small conical

☐ medium conical

☐ flat

☐ hooked

☐ long curved

☐ crossed

Probable name :

Diagram :

Date :

Tail :

☐ squared

☐ forked

☐ stepped

Social :

☐ solitary

☐ in group

Color(s) :

© All rights reserved — Nature logbook by Mickaël Nicotera

Bird observation sheet

Beak:

- ☐ small conical
- ☐ medium conical
- ☐ flat
- ☐ hooked
- ☐ long curved
- ☐ crossed

Probable name:

Diagram:

Date:

Tail:
- ☐ squared
- ☐ forked
- ☐ stepped

Social:
- ☐ solitary
- ☐ in group

Color(s):

© All rights reserved – Nature logbook by Mickaël Nicotera

Bird observation sheet

Beak :

☐ small conical

☐ medium conical

☐ flat

☐ hooked

☐ long curved

☐ crossed

Probable name :

Diagram :

Date :

Tail :

☐ squared

☐ forked

☐ stepped

Social :

☐ solitary

☐ in group

Color(s) :

© All rights reserved – Nature logbook by Mickaël Nicotera

Bird observation sheet

Beak:

☐ small conical

☐ medium conical

☐ flat

☐ hooked

☐ long curved

☐ crossed

Probable name:

Diagram:

Date:

Tail:

☐ squared

☐ forked

☐ stepped

Social:

☐ solitary

☐ in group

Color(s):

© All rights reserved – Nature logbook by Mickaël Nicotera

Bird observation sheet

Beak :

- ☐ small conical
- ☐ medium conical
- ☐ flat
- ☐ hooked
- ☐ long curved
- ☐ crossed

Probable name :

Diagram :

Date :

Tail :

- ☐ squared
- ☐ forked
- ☐ stepped

Social :

- ☐ solitary
- ☐ in group

Color(s) :

© All rights reserved – Nature logbook by Mickaël Nicotera

Fungi

Fungi are a particular group of organisms. They are neither plants nor animals.

They grow from spores and are in symbiotic relationship with tree roots.

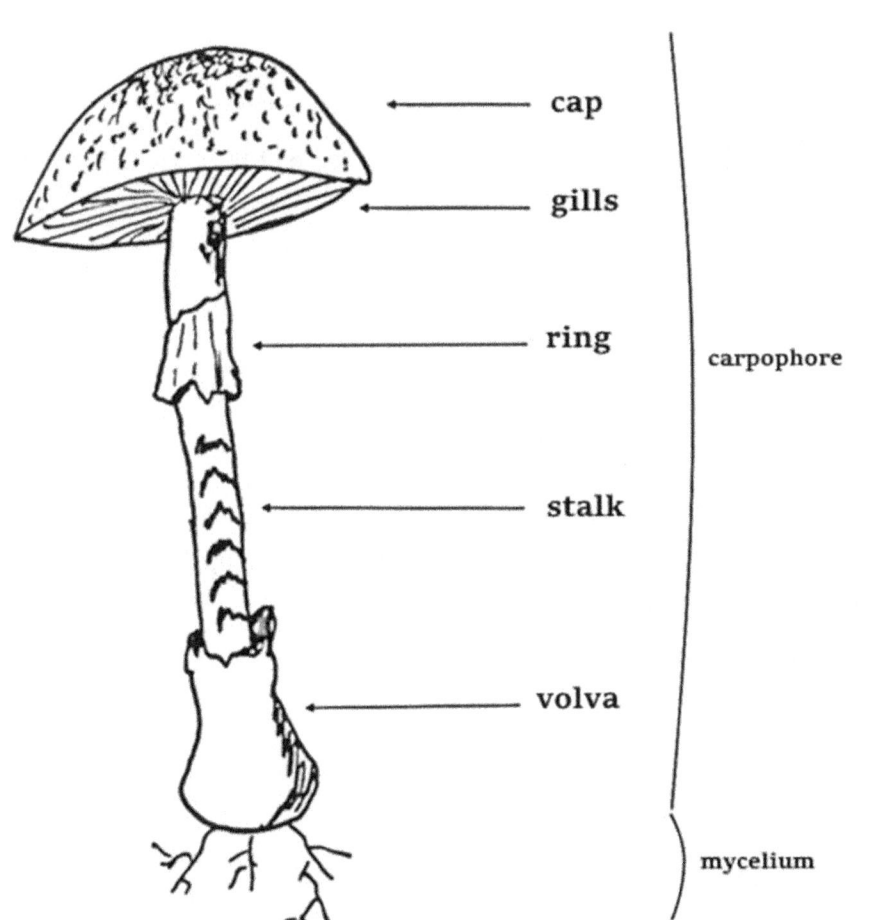

Common fungi

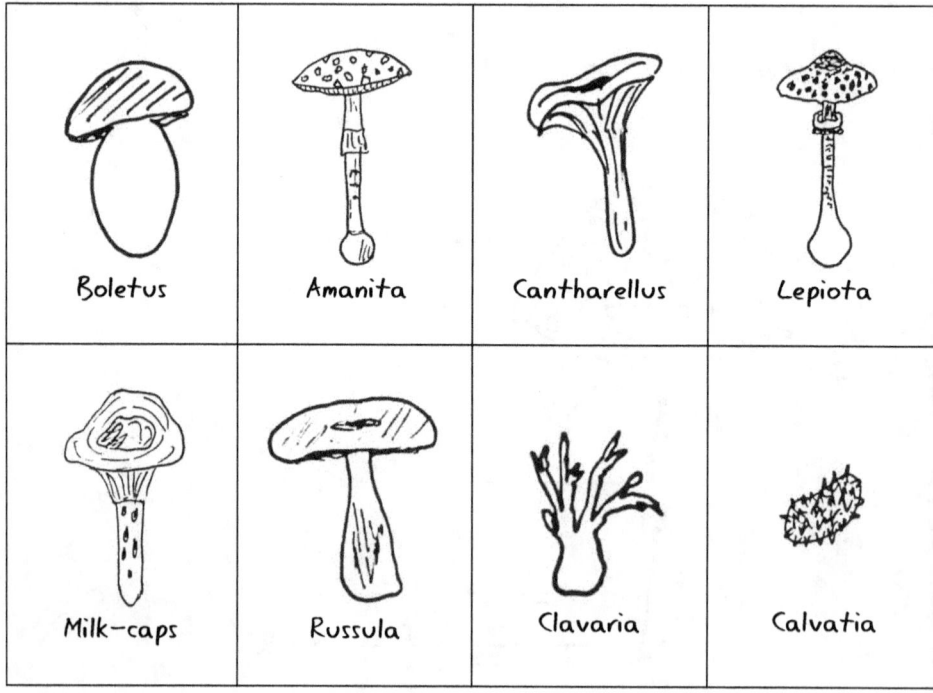

Fungi observation sheet

Stipe:
- ☐ equal
- ☐ fusoid
- ☐ bulbous
- ☐ clavate
- ☐ ventricose
- ☐ radicating

Probable name :

Diagram :

Date :

Cap :
- ☐ convex
- ☐ knobbed
- ☐ conical
- ☐ depressed

Color(s) :

© All rights reserved — Nature logbook by Mickaël Nicotera

Fungi observation sheet

Stipe:
- ☐ equal
- ☐ fusoid
- ☐ bulbous
- ☐ clavate
- ☐ ventricose
- ☐ radicating

Probable name:

Diagram:

Date:

Cap:
- ☐ convex
- ☐ knobbed
- ☐ conical
- ☐ depressed

Color(s):

Fungi observation sheet

Stipe:
- ☐ equal
- ☐ fusoid
- ☐ bulbous
- ☐ clavate
- ☐ ventricose
- ☐ radicating

Probable name :

Diagram :

Date :

Cap :
- ☐ convex
- ☐ knobbed
- ☐ conical
- ☐ depressed

Color(s) :

© All rights reserved – Nature logbook by Mickaël Nicotera

Fungi observation sheet

Stipe:
- ☐ equal
- ☐ fusoid
- ☐ bulbous
- ☐ clavate
- ☐ ventricose
- ☐ radicating

Probable name :

Diagram :

Date :

Cap :
- ☐ convex
- ☐ knobbed
- ☐ conical
- ☐ depressed

Color(s) :

© All rights reserved – Nature logbook by Mickaël Nicotera

Fungi observation sheet

Stipe:
- ☐ equal
- ☐ fusoid
- ☐ bulbous
- ☐ clavate
- ☐ ventricose
- ☐ radicating

Probable name :

Diagram :

Date :

Cap :
- ☐ convex
- ☐ knobbed
- ☐ conical
- ☐ depressed

Color(s) :

© All rights reserved – Nature logbook by Mickaël Nicotera

Flowers

Flowers are the reproductive organ of many plants. They may be solitary or part of an inflorescence. Their colors and shapes combine in countless ways.

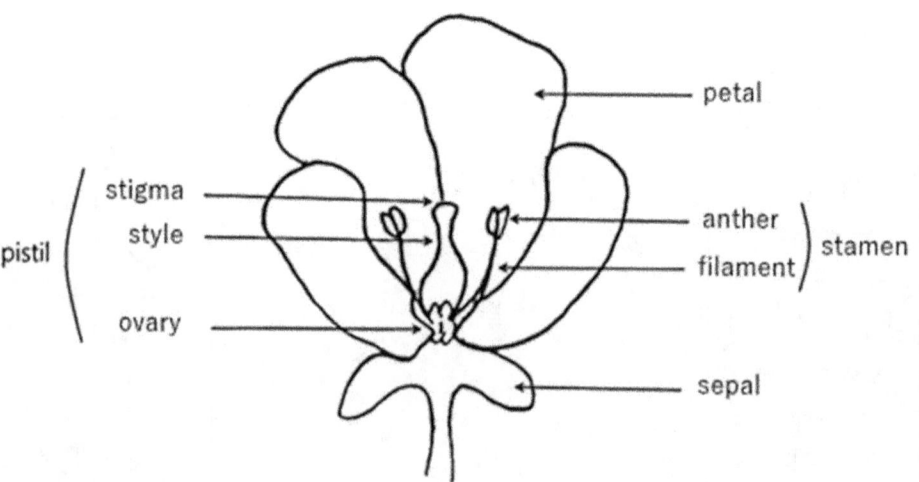

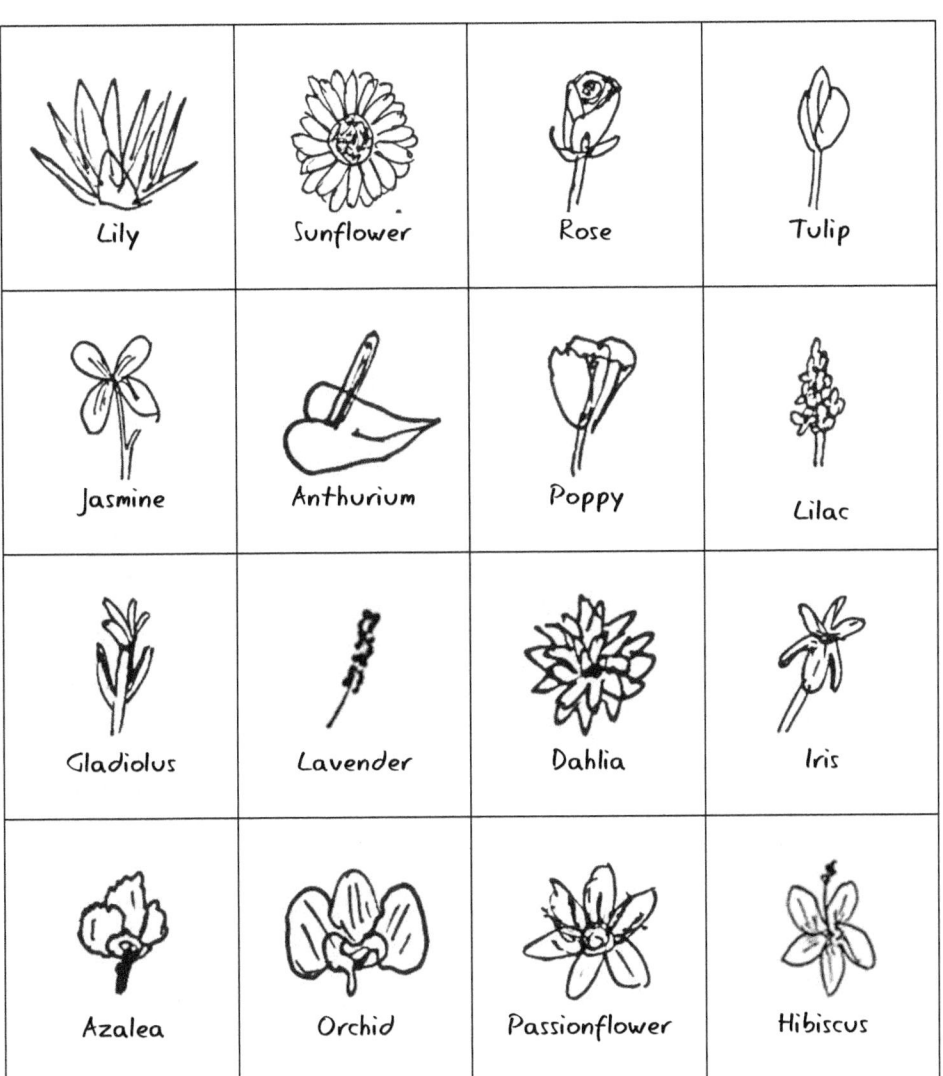

Flower observation sheet

Probable name :

Date :

☐ separate petals
☐ fused petals
☐ dense petals
☐ bell-shaped
☐ tube-shaped
☐ papilionaceous
☐ personate

Diagram :

☐ solitary
☐ inflorescence

Inflorescence

☐ cyme
☐ spike
☐ racene
☐ panicle
☐ corymb
☐ umbel
☐ capitulum

Color(s) :

Flower observation sheet

	Probable name :	Date :
☐ separate petals ☐ fused petals ☐ dense petals ☐ bell-shaped ☐ tube-shaped ☐ papilionaceous ☐ personate	Diagram : ☐ solitary ☐ inflorescence	Inflorescence ☐ cyme ☐ spike ☐ racene ☐ panicle ☐ corymb ☐ umbel ☐ capitulum

Color(s) :

© All rights reserved – Nature logbook by Mickaël Nicotera

Flower observation sheet

Probable name :

Date :

- ☐ separate petals
- ☐ fused petals
- ☐ dense petals
- ☐ bell-shaped
- ☐ tube-shaped
- ☐ papilionaceous
- ☐ personate

Diagram :

☐ solitary
☐ inflorescence

Inflorescence

- ☐ cyme
- ☐ spike
- ☐ racene
- ☐ panicle
- ☐ corymb
- ☐ umbel
- ☐ capitulum

Color(s) :

© All rights reserved – Nature logbook by Mickaël Nicotera

Flower observation sheet

	Probable name :	Date :
☐ separate petals ☐ fused petals ☐ dense petals ☐ bell-shaped ☐ tube-shaped ☐ papilionaceous ☐ personate	Diagram : ☐ solitary ☐ inflorescence	Inflorescence ☐ cyme ☐ spike ☐ racene ☐ panicle ☐ corymb ☐ umbel ☐ capitulum

Color(s) :

Flower observation sheet

Probable name :

Date :

- ☐ separate petals
- ☐ fused petals
- ☐ dense petals
- ☐ bell-shaped
- ☐ tube-shaped
- ☐ papilionaceous
- ☐ personate

Diagram :

☐ solitary
☐ inflorescence

Inflorescence

- ☐ cyme
- ☐ spike
- ☐ racene
- ☐ panicle
- ☐ corymb
- ☐ umbel
- ☐ capitulum

Color(s) :

Insects

All insects have three body regions: the head, the thorax and the abdomen. Three pairs of legs are attached to the thorax and two antennae are attached to the head.

Some insects can have wings (two or four) to fly.

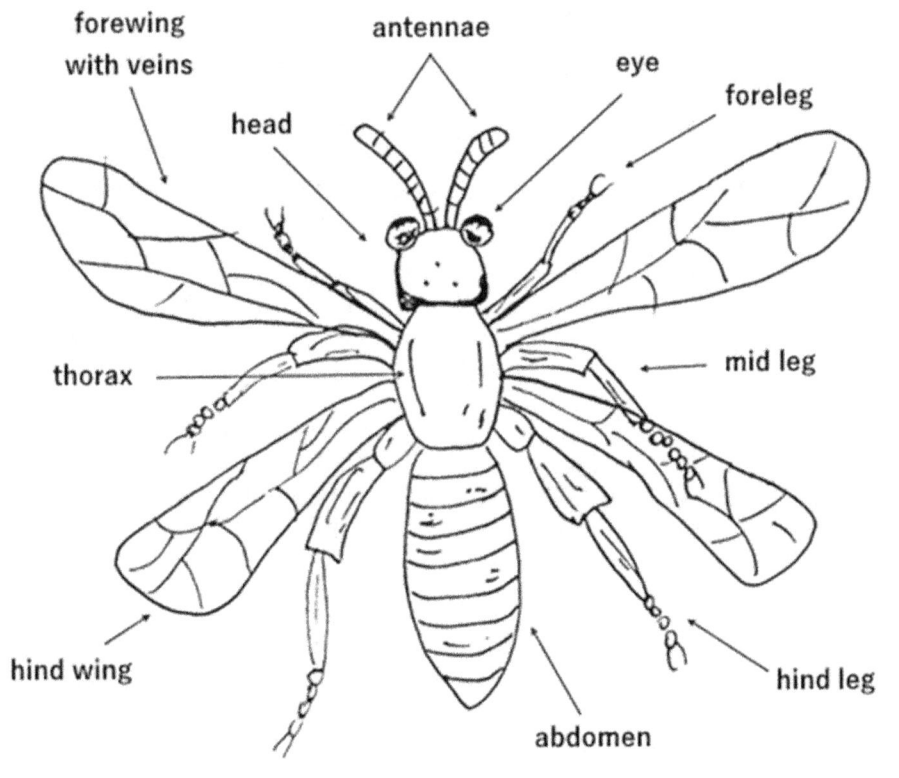

Common insects

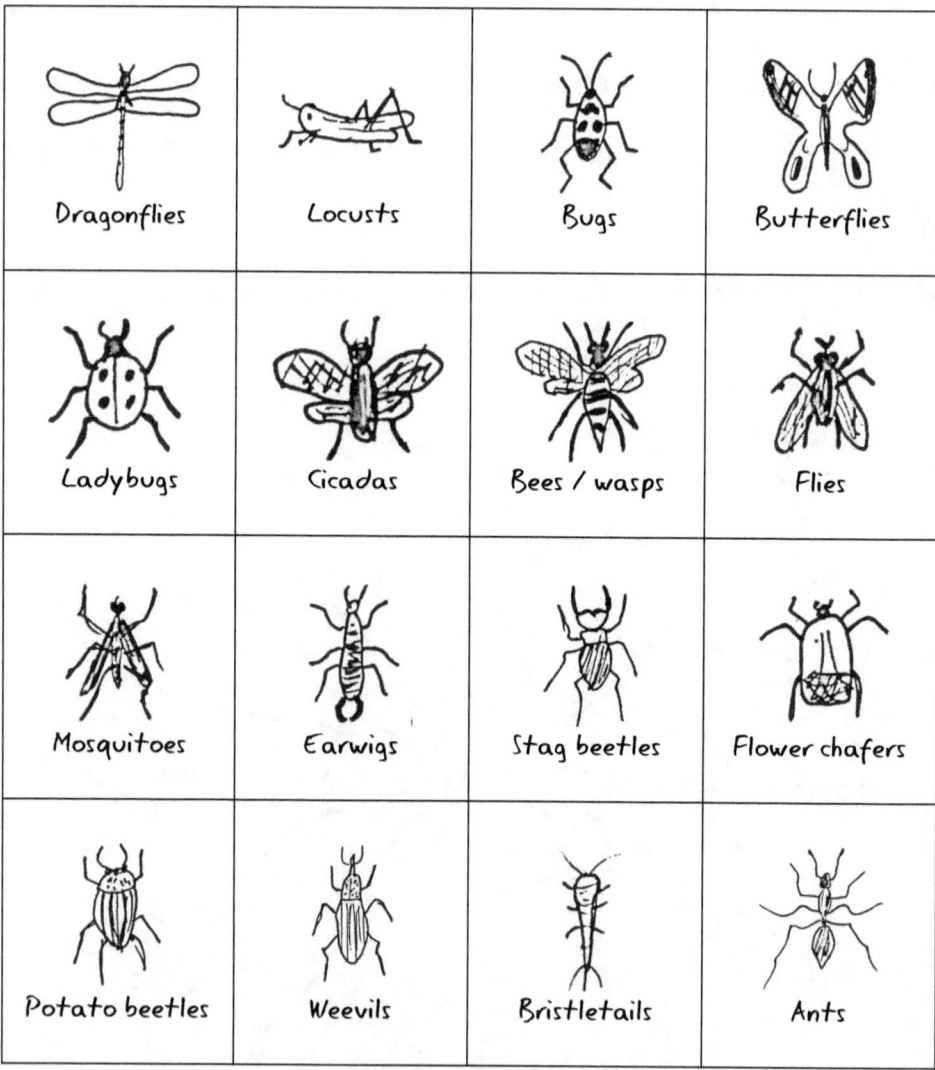

Insect observation sheet

Antennae:
- ☐ aristate
- ☐ lamellate
- ☐ geniculate
- ☐ setaceous
- ☐ clavate
- ☐ plumose

Probable name:

Diagram:

- ☐ solitary
- ☐ in group

Date:

- ☐ winged
- ☐ elytra
- ☐ fringed
- ☐ scaly
- ☐ frenulum
- ☐ halteres

Color(s):

Insect observation sheet

Antennae :

☐ aristate

☐ lamellate

☐ geniculate

☐ setaceous

☐ clavate

☐ plumose

Probable name :

Diagram :

☐ solitary
☐ in group

Date :

☐ winged

☐ elytra

☐ fringed

☐ scaly

☐ frenulum

☐ halteres

Color(s) :

© All rights reserved – Nature logbook by Mickaël Nicotera

Insect observation sheet

Antennae :
- ☐ aristate
- ☐ lamellate
- ☐ geniculate
- ☐ setaceous
- ☐ clavate
- ☐ plumose

Probable name :

Diagram :

- ☐ solitary
- ☐ in group

Date :

- ☐ winged
- ☐ elytra
- ☐ fringed
- ☐ scaly
- ☐ frenulum
- ☐ halteres

Color(s) :

© All rights reserved – Nature logbook by Mickaël Nicotera

Insect observation sheet

Antennae:
- ☐ aristate
- ☐ lamellate
- ☐ geniculate
- ☐ setaceous
- ☐ clavate
- ☐ plumose

Probable name:

Diagram:

- ☐ solitary
- ☐ in group

Date:

- ☐ winged
- ☐ elytra
- ☐ fringed
- ☐ scaly
- ☐ frenulum
- ☐ halteres

Color(s):

© All rights reserved – Nature logbook by Mickaël Nicotera

Insect observation sheet

Antennae :

☐ aristate

☐ lamellate

☐ geniculate

☐ setaceous

☐ clavate

☐ plumose

Probable name :

Diagram :

☐ solitary
☐ in group

Date :

☐ winged

☐ elytra

☐ fringed

☐ scaly

☐ frenulum

☐ halteres

Color(s) :

Printed by Lulu.com
January 2016
ISBN : 978-1-326-54332-7

www.ingramcontent.com/pod-product-compliance
Lightning Source LLC
Chambersburg PA
CBHW072305170526
45158CB00003BA/1198